Sponsored Technology Workshops 2023

AA008707

Held at the 2023 AIChE Annual Meeting

Orlando, Florida, USA
5-10 November 2023

ISBN: 978-1-7138-9324-0

Printed from e-media with permission by:

Curran Associates, Inc.
57 Morehouse Lane
Red Hook, NY 12571

Some format issues inherent in the e-media version may also appear in this print version.

Copyright© (2023) by AIChE
All rights reserved.

Printed with permission by Curran Associates, Inc. (2024)

For permission requests, please contact AIChE
at the address below.

AIChE
120 Wall Street, FL 23
New York, NY 10005-4020

Phone: (800) 242-4363
Fax: (203) 775-5177

www.aiche.org

Additional copies of this publication are available from:

Curran Associates, Inc.
57 Morehouse Lane
Red Hook, NY 12571 USA
Phone: 845-758-0400
Fax: 845-758-2633
Email: curran@proceedings.com
Web: www.proceedings.com

TABLE OF CONTENTS

714a Combining Machine Learning and Simulation for the Virtual Optimization of Bulk Solids Handling Processes at the Micro and Macro Scales 1
Stefan Pantaleev

716a From Efficient 3D Simulations to Digital Twins and Everything In-Between: Modeling for the Process and Pharma Industries 2
Hossam Metwally

7720a Coprocessing to Optimize Bioavailability 3
San Kiang

717a Unlock Accelerated Innovation and Cost Savings in Life Sciences and Energy Sectors with Cloud-Based High-Performance Computing 4
Dylan Collie

718a PANEL DISCUSSION - Application of Modeling, Simulation, and Advanced Control in Process Development and Manufacturing of Chemical Entities 6
Neda Nazemifard

719 Multi-Scale Computational Chemistry for Drug Product Formulation & Development 7
Jan-Willem Handgraaf

Author Index

714a Combining Machine Learning and Simulation for the Virtual Optimization of Bulk Solids Handling Processes at the Micro and Macro Scales

Stefan Pantaleev

Virtual process optimization is a key component of the digital transformation strategy in a wide range of bulk solids handling industries with the traditional physics-based simulation approach already reducing manufacturing costs and time to market for new products. Machine learning can be used to substantially increase the benefit of physics-based simulations by reducing computational expense and maximizing learning. Join this workshop to discover how Altair simulation and machine learning tools such as EDEM and HyperStudy can be used to efficiently optimize bulk solids handling processes at both the particle and equipment scales.

716a From Efficient 3D Simulations to Digital Twins and Everything in-between: Modeling for the Process and Pharma Industries

Hossam Metwally

In the journey towards digitalization, generating simulation-based data is an integral part. Generating enough data requires simulation to be fast, and the simulation process to be efficient. In this seminar, recent advances in the GPU solver will be highlighted that has shown simulations to be up to 8x when compared with existing simulation best practices. Additionally, examples of streamlining the simulation workflow will be shown which enables the creation of Design of Experiments (DOE) further opening the door for optimization and meta/surrogate/reduced-order model creation. Finally, and additionally, advances in system-level modeling will also be highlighted where field/experimental data can be fused with simulation-based data to provide for an accurate digital representation of the asset/system and bring the power of simulation to the hands of non-experts for operational decision-making through virtual sensors.

This will be a curated networking discussion for only a limited portion of the audience, so make sure to ***RSVP*** *here.* ***Lunch will be provided!***

7720a Coprocessing to Optimize Bioavailability

San Kiang

As therapies become more complex, so do the obstacles that can interfere with their delivery. A robust solution to inadequate drug solubility, powder flow, and bioavailability, continuous co-processing is a platform that can be applied across various molecule classes and formulation types. Utilizing both classical and innovative techniques, co-processing has the added benefit of expediting manufacture of materials needed for toxicology screens and clinical studies. Additionally, co-processing assures the scale up of many energetic and sensitive processes, thus allowing for a safe and rugged technical transfer. This technical showcase will highlight how the co-processing platform and methodology behind it can be your next solution to drug development and delivery.

Learning Objectives:

- *Understand what is Co-processing, the latest technologies used in the technique, as well as the most up-to-date regulatory status*
- *Discover the ongoing industry efforts to promote this technology*
- *Learn some of the common techniques used on the co-processing technology platform*
- *Know some example(s) of how the use of co-processing can improve powder flow and density*
- *Know some example(s) of how the use of co-processing can improve formulation process ruggedness*

717a Unlock Accelerated Innovation and Cost Savings in Life Sciences and Energy Sectors with Cloud-Based High-Performance Computing

Dylan Collie

Computational simulation continues to play an increasingly critical role in the understanding and development of products and processes within the life sciences and energy sectors.

Huge strides and advancements in high performance computing in the cloud have unlocked acceleration opportunities for computational fluid dynamics, computational chemistry, and molecular dynamic simulations to discover new drugs and materials.

Join this lunch presentation to learn about the latest technologies scientists, engineers, and R&D teams are leveraging to save costs and accelerate the rate of innovation, commercializing technology more efficiently.

Learn how to:

1. Streamline Workflows: Learn how easy it is to access a diverse range of over 100+ HPC architectures, from NVIDIA GPUs to custom software toolkits like Python and MATLAB.
2. Maximize ROI: Hear success stories from industry leaders on how they've slashed computational costs by over 50% and time-to-insight by 30% using Rescale.
3. Secure Your Data: Understand how our platform meets stringent security and compliance requirements while enabling enhanced collaboration and data sharing.

Rescale's Multi-Cloud Environment gives you access to 100+ HPC architectures, including NVIDIA GPUs, for optimal software workload performance across a range of scientific and engineering applications.

Rescale is built for any software (commercial, open source, or custom) and physics; from Molecular Dynamics to Density Functional Theory, optimization will speed up your time to answer.

You can also decrease the training time for your data science teams with Python, R, TensorFlow, and AI accelerators optimized on bespoke hardware.

In this workshop, we'll also provide a demo walkthrough of running computational chemistry codes such as GROMACS, LAMMPS or Quantum Espresso on the Rescale Platform.

"Through Rescale, we were able to draw results more than 2x faster than our existing workflow, which helped us make better business decisions."

- *Jerry Maeng Managing Director, AZothBio*

"Bionano's use of Azure on Rescale reduces compute cost for genetic disease analysis by more than 50%, compute time by 30%."

- *Mark Oldakowski, COO, Bionano*

"Rescale simplifies high performance computing, which helps us improve our models and make new discoveries faster. Ultimately it's about commercializing our technology faster so we can

accelerate global decarbonization."

- *Seonghoon Woo, CEO, Amogy*

718a PANEL DISCUSSION - Application of modeling, simulation, and advanced control in process development and manufacturing of chemical entities

Neda Nazemifard

Process modeling and simulations of all types; mechanistic, AI/ML, hybrid, system models as well as advanced control strategies have been an integral part of process development and manufacturing across the chemical industry in recent years. It is great to see that the pharmaceutical industry is catching up with digitalization principles and embracing advanced modeling, simulations, digital twins, and control strategies in achieving Industry 5.0 goals. In this workshop, we are seeking input from scientists and engineers across the pharmaceutical industry and larger chemical industry to share their knowledge and experiences in the application of modeling and advanced control during the lifecycle of the process from design, understanding, optimization, validation to tech transfer to manufacturing. We have invited a panel consisting of experienced SME's across industry and academia to drive this discussion. The topics of discussion are around but not limited to:

Use Cases and Success Stories: Share specific case studies or success stories where digital twins and advanced control have made a significant impact on process development, optimization, or manufacturing efficiency.

Challenges and Bottlenecks: Discuss the challenges faced by the pharmaceutical industry in adopting these technologies, including regulatory hurdles, data integration issues, and cultural resistance within organizations.

Tech Transfer and Scaling Up: Share experiences and strategies for successfully transferring technology and processes from R&D to manufacturing while maintaining consistency and efficiency.

Data Management and Security: Address the importance of data management, security, and privacy in the context of digital twins and advanced control systems,

Collaboration between Industry and Academia: Discuss the role of academia in advancing these technologies and how industry-academia collaboration can accelerate innovation.

719 Multi-scale computational chemistry for drug product formulation & development

Jan-Willem Handgraaf

A key challenge in commercializing new drug products is the transition from early-stage discovery in the laboratory to scaling up in the engineering stage. This is primarily because of the interdependency of various components in a formulation such as solubility, hydrophobicity, phase-interaction, stability, etc.
A standard drug product development program can be slowed down until the right choice of an optimum combination can be found. This requires expensive and time-consuming experimentation to screen a large set of candidates.

In this event, we will present computational solutions to these challenges. We will discuss several workflows to evaluate these behaviors such as

- Partitioning of chemicals in emulsions
- Protein formulations, aggregation & stability
- Phase-behavior& micro-emulsion propensity calculations

These are applied in the following use cases

1. Vaccine stability and shelf-life
2. Medical nutrition
3. Lyophilization

We will demonstrate how these methods and tools can ease the pain of transitioning from early-stage discovery to scale-up stage. We will show cases where it has led to significant time & cost savings through reduced trial and error and higher confidence during the following scale-up stage.

References
https://plm.sw.siemens.com/en-US/simcenter/fluids-thermal-simulation/culgi/
Computational solutions for thermodynamic and rheological properties

AUTHOR INDEX

Collie, Dylan....................4
Handgraaf, Jan-Willem....................7
Kiang, San....................3
Metwally, Hossam....................2
Nazemifard, Neda....................6
Pantaleev, Stefan....................1

NOTES

NOTES

NOTES

NOTES

NOTES